Applying the Standards:
STEM
Kindergarten

Credits
Content Editor: Jennifer B. Stith
Copy Editor: Karen Seberg

Visit carsondellosa.com for correlations to Common Core, state, national, and Canadian provincial standards.

Carson-Dellosa Publishing, LLC
PO Box 35665
Greensboro, NC 27425 USA
carsondellosa.com

ISBN 978-1-4838-1566-4
02-278151151

Table of Contents

Introduction

STEM education is a growing force in today's classroom. Exposure to science, technology, engineering, and math is important in twenty-first century learning as it allows students to succeed in higher education as well as a variety of careers.

While it can come in many forms, STEM education is most often presented as an engaging task that asks students to solve a problem. Additionally, creativity, collaboration, communication, and critical thinking are integral to every task. STEM projects are authentic learning tasks that guide students to address a variety of science and math standards. Also, students strengthen English Language Arts skills by recording notes and written reflections throughout the process.

In this book, students are asked to complete a range of tasks with limited resources. Materials for each task are limited to common household objects. Students are guided through each task by the steps of the engineering design process to provide a framework through which students can grow their comfort level and independently complete tasks.

Use the included rubric to guide assessment of student responses and further plan any necessary remediation. Confidence in STEM tasks will help students succeed in their school years and beyond.

Student Roles

Student collaboration is an important component of STEM learning. Encourage collaboration by having students complete tasks in groups. Teach students to communicate openly, support each other, and respect the contributions of all members. Keep in mind that collaborative grouping across achievement levels can provide benefits for all students as they pool various perspectives and experiences toward a group goal.

Consider assigning formal roles to students in each group. This will simplify the collaborative tasks needed to get a project done and done well. The basic roles of group structure are as follows:

- The *captain* leads and guides other students in their roles.

- The *guide* walks the team through the steps, keeps track of time, and encourages the team to try again.

- The *materials manager* gathers, organizes, and guides the use of materials.

- The *reporter* records the team's thoughts and reports on the final project to the class.

STEM Performance Rubric

Use this rubric as a guide for assessing students' project management skills. It can also be offered to students as a tool to show your expectations and scoring. Note: Some items may not apply to each project.

4	_____ Asks or identifies comprehensive high-level questions _____ Makes valid, nontrivial inferences based on evidence in the text _____ Uses an appropriate, complete strategy to solve the problem _____ Skillfully justifies the solution and strategy used _____ Offers insightful reasoning and strong evidence of critical thinking _____ Collaborates with others in each stage of the process _____ Effectively evaluates and organizes information and outcomes
3	_____ Asks or identifies ample high-level questions _____ Exhibits effective imagination and creativity _____ Uses an appropriate but incomplete strategy to solve the problem _____ Justifies answer and strategy used _____ Offers sufficient reasoning and evidence of critical thinking _____ Collaborates with others in most stages of the process _____ Evaluates and organizes some information or outcomes
2	_____ Asks or identifies a few related questions _____ Exhibits little imagination and creativity _____ Uses an inappropriate or unclear strategy for solving the problem _____ Attempts to justify answers and strategy used _____ Demonstrates some evidence of critical thinking _____ Collaborates with others if prompted _____ Can evaluate and organize simple information and outcomes
1	_____ Is unable to ask or identify pertinent questions _____ Does not exhibit adequate imagination and creativity _____ Uses no strategy or plan for solving the problem _____ Does not or cannot justify answer or strategy used _____ Demonstrates limited or no evidence of critical thinking _____ Does not collaborate with others _____ Cannot evaluate or organize information or outcomes

Name _____

Read the task. Then, follow the directions to complete the task.

Color Creations

Use water and drops of food coloring to make all of the colors of a rainbow—red, orange, yellow, green, blue, indigo, and violet.

Materials

water
clear plastic cups

red, blue, and yellow
liquid food coloring
plastic spoons

🔎 Ask

What do you know? What do you need to know to get started?

_____

_____

💭 Imagine

What could you do?

_____

_____

📓 Plan

Choose an idea. Draw a model.

Plan

What are your steps? Use your model to guide your plan.

_____

_____

_____

Create

Follow your plan. What is working? Do you need to try something else?

_____

_____

Improve

How could you make it better?

_____

_____

Communicate

How well did it work? Is the problem solved?

_____

_____

Reflect

What colors were easiest to make? Hardest? Why?

_____

_____

Name _____

Read the task. Then, follow the directions to complete the task.

3, 2, 1, Blast Off!

Create a paper rocket that when launched will travel at least 5 feet (1.5 m).

Materials

copy paper	tape
plastic drinking straws	scissors

Ask

What do you know? What do you need to know to get started?

_____

_____

Imagine

What could you do?

_____

_____

Plan

Choose an idea. Draw a model.

📓 Plan

What are your steps? Use your model to guide your plan.

_____

_____

_____

🛠 Create

Follow your plan. What is working? Do you need to try something else?

_____

_____

🔄 Improve

How could you make it better?

_____

_____

💬 Communicate

How well did it work? Is the problem solved?

_____

_____

🔆 Reflect

How was the straw important to your design?

_____

_____

Name _____

Read the task. Then, follow the directions to complete the task.

Alien Animals

Make an animal that can stand on 3 legs and is at least 4 cubes long. You may add other body parts such as fur, eyes, ears, arms, and antennae.

Materials

play dough
golf tees
large brass paper
    fasteners
cotton swabs
toothpicks

self-adhesive wiggly
    eyes
chenille stems
buttons
linking cubes

Ask

What do you know? What do you need to know to get started?

_____

_____

Imagine

What could you do?

_____

_____

Plan

Choose an idea. Draw a model.

📓 Plan

What are your steps? Use your model to guide your plan.

_____

_____

_____

🔧 Create

Follow your plan. What is working? Do you need to try something else?

_____

_____

🔄 Improve

How could you make it better?

_____

_____

💬 Communicate

How well did it work? Is the problem solved?

_____

_____

☀️ Reflect

How did the placement of the animal's legs help it stand up?

_____

_____

Name _____

Read the task. Then, follow the directions to complete the task.

Big Bubble Wand

Use chenille stems to make a bubble wand that will make the biggest bubble possible.

Materials

chenille stems   shallow pie pan
bubble solution or
water and dish soap

Ask

What do you know? What do you need to know to get started?

_____

_____

Imagine

What could you do?

_____

_____

Plan

Choose an idea. Draw a model.

📓 Plan

What are your steps? Use your model to guide your plan.

_____

_____

_____

🛠️ Create

Follow your plan. What is working? Do you need to try something else?

_____

_____

🔄 Improve

How could you make it better?

_____

_____

💬 Communicate

How well did it work? Is the problem solved?

_____

_____

☀️ Reflect

How did the size of the wand affect the size of the bubble?

_____

_____

Name _____

Read the task. Then, follow the directions to complete the task.

Bottle Cap Balance

Balance a bottle cap on only 1 straw. Your bottle cap must stay balanced for 1 minute.

Materials

plastic bottle cap	scissors
plastic drinking straw	timer

🔑 Ask

What do you know? What do you need to know to get started?

_____

_____

💭 Imagine

What could you do?

_____

_____

📝 Plan

Choose an idea. Draw a model.

[blank box]

✏️ Plan

What are your steps? Use your model to guide your plan.

_____

_____

_____

🔧 Create

Follow your plan. What is working? Do you need to try something else?

_____

_____

🔄 Improve

How could you make it better?

_____

_____

💬 Communicate

How well did it work? Is the problem solved?

_____

_____

☀️ Reflect

How was the number of straw pieces used important to balancing the bottle cap?

_____

_____

Name _____

Read the task. Then, follow the directions to complete the task.

Really Wheel-y

Build a wheel using gumdrops, wooden skewers, and toothpicks. The wheel should roll straight for at least 12 inches (30 cm).

Materials

gumdrops   toothpicks
wooden skewers   yardstick (meterstick)

Caution: Ensure that students are properly supervised while using sharp objects.

Caution: Before beginning any food activity, ask families' permission and inquire about students' food allergies and religious or other food restrictions.

Ask

What do you know? What do you need to know to get started?

_____

_____

Imagine

What could you do?

_____

_____

Plan

Choose an idea. Draw a model.

📝 Plan

What are your steps? Use your model to guide your plan.

_____

_____

_____

✂️ Create

Follow your plan. What is working? Do you need to try something else?

_____

_____

🔄 Improve

How could you make it better?

_____

_____

💬 Communicate

How well did it work? Is the problem solved?

_____

_____

🌟 Reflect

How did the amount of materials used affect how far and how fast the wheel rolled?

_____

_____

Name _____

Read the task. Then, follow the directions to complete the task.

Scent Sleuths: Exploring the Sense of Smell

Imagine that you are having a picnic. While you are on a walk, an animal takes a bite out of your lemon pie. Find out who did it using only your sense of smell.

Materials

opaque containers with small holes in the lids, each labeled with an animal name

a variety of food scents from extracts or actual pieces of food

Caution: Before beginning any food activity, ask families' permission and inquire about students' food allergies and religious or other food restrictions.

Caution: Before beginning this activity, ask families' permission and inquire about students' scent sensitivities and/ or allergies.

Ask

What do you know? What do you need to know to get started?

_____

_____

Imagine

What could you do?

_____

_____

Plan

Choose an idea. Draw a model.

📝 Plan

What are your steps? Use your model to guide your plan.

_____

_____

_____

🛠 Create

Follow your plan. What is working? Do you need to try something else?

_____

_____

🔄 Improve

How could you make it better?

_____

_____

💬 Communicate

How well did it work? Is the problem solved?

_____

_____

🌟 Reflect

How does your sense of smell help you understand the world around you?

_____

_____

Name _____

Read the task. Then, follow the directions to complete the task.

Handy Art: Exploring the Sense of Touch

Make a piece of touchable art. Others must be able to "see" the art using only their hands.

Materials

drawing paper	art sand
poster board	pom-poms
construction paper	beads
tempera paint	chenille stems
paintbrushes	buttons
kosher salt	glue
table salt	scissors

Caution: Before beginning any food activity, ask families' permission and inquire about students' food allergies and religious or other food restrictions.

Ask

What do you know? What do you need to know to get started?

_____

_____

Imagine

What could you do?

_____

_____

Plan

Choose an idea. Draw a model.

Plan

What are your steps? Use your model to guide your plan.

_____

_____

_____

Create

Follow your plan. What is working? Do you need to try something else?

_____

_____

Improve

How could you make it better?

_____

_____

Communicate

How well did it work? Is the problem solved?

_____

_____

Reflect

What was the most important thing to do when creating your piece of art? Why?

_____

_____

 © Carson-Dellosa · CD-104846 · Applying the Standards: STEM

Name _____

Read the task. Then, follow the directions to complete the task.

Drumbeat Band: Exploring the Sense of Hearing

Make 2 drums. The drums must make different sounds when played.

Materials

empty cylinder-
   shaped containers
   such as for rolled
   oats, coffee,
   powdered drinks, etc.
aluminum foil
paper
plastic wrap

waxed paper
cellophane
rubber bands
string
tape
glue
scissors
unsharpened pencils

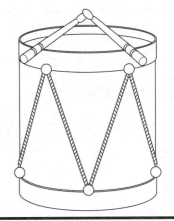

⚡ Ask

What do you know? What do you need to know to get started?

_____

_____

💭 Imagine

What could you do?

_____

_____

📝 Plan

Choose an idea. Draw a model.

📓 Plan

What are your steps? Use your model to guide your plan.

_____

_____

_____

🛠 Create

Follow your plan. What is working? Do you need to try something else?

_____

_____

🔄 Improve

How could you make it better?

_____

_____

💬 Communicate

How well did it work? Is the problem solved?

_____

_____

🌟 Reflect

How did the covering used for the top of the drum affect the sound the drum made?

_____

_____

Name _____

Read the task. Then, follow the directions to complete the task.

Count and Stack

Pick a number card from 1 to 20. Build the tallest tower you can with that number of cups. The tower must stand for 1 minute.

Materials

set of number cards   20 small, clear plastic
  1–20   cups
                        timer

Ask

What do you know? What do you need to know to get started?

_____

_____

Imagine

What could you do?

_____

_____

Plan

Choose an idea. Draw a model.

📝 Plan

What are your steps? Use your model to guide your plan.

_____

_____

_____

🛠 Create

Follow your plan. What is working? Do you need to try something else?

_____

_____

🔄 Improve

How could you make it better?

_____

_____

💬 Communicate

How well did it work? Is the problem solved?

_____

_____

☀ Reflect

How did the number you picked affect the height of your tower? How did the number affect if your tower could stand for 1 minute?

_____

_____

Name _____

Read the task. Then, follow the directions to complete the task.

Icy Escape

Using only your body, remove an object from a block of ice.

Materials

large ice cube with a
    small object such
    as a coin, button,
    or plastic toy frozen
    inside

plastic tray or baking
    sheet

Ask

What do you know? What do you need to know to get started?

_____

_____

Imagine

What could you do?

_____

_____

Plan

Choose an idea. Draw a model.

📝 Plan

What are your steps? Use your model to guide your plan.

_____

_____

_____

🛠 Create

Follow your plan. What is working? Do you need to try something else?

_____

_____

🔄 Improve

How could you make it better?

_____

_____

💬 Communicate

How well did it work? Is the problem solved?

_____

_____

☀ Reflect

Think about what you know about ice and water. How did what you know help you remove the object from the ice?

_____

_____

Name _____

Read the task. Then, follow the directions to complete the task.

Sail Away

Create a sail to help move a "boat" across water. Your breath can be the wind!

Materials

large container of
    water
small, lightweight
    plastic bowl
wooden craft sticks
wooden skewers
toothpicks

copy paper
parchment paper
play dough
glue
tape
scissors

Caution: Ensure that students are properly supervised while using sharp objects.

 Ask

What do you know? What do you need to know to get started?

_____

_____

 Imagine

What could you do?

_____

_____

Plan

Choose an idea. Draw a model.

📝 Plan

What are your steps? Use your model to guide your plan.

_____

_____

_____

🛠 Create

Follow your plan. What is working? Do you need to try something else?

_____

_____

🔄 Improve

How could you make it better?

_____

_____

💬 Communicate

How well did it work? Is the problem solved?

_____

_____

🌟 Reflect

Explain how the "wind" helped move your boat.

_____

_____

Name _____

Read the task. Then, follow the directions to complete the task.

Stop That Marble: Friction

Find the best material to stop a falling marble. The marble must be dropped from 12 inches (30 cm) above a container.

Materials

marble
clear plastic bottle
   with the top cut off
measuring cups
rice
feathers
unpopped popcorn

popped popcorn
sand
shredded paper
gelatin dessert or
   vegetarian
   alternative
ruler

Caution: Before beginning any food activity, ask families' permission and inquire about students' food allergies and religious or other food restrictions.

Ask

What do you know? What do you need to know to get started?

_____

_____

Imagine

What could you do?

_____

_____

Plan

Choose an idea. Draw a model.

📝 Plan

What are your steps? Use your model to guide your plan.

_____

_____

_____

🔧 Create

Follow your plan. What is working? Do you need to try something else?

_____

_____

🔄 Improve

How could you make it better?

_____

_____

💬 Communicate

How well did it work? Is the problem solved?

_____

_____

☀ Reflect

What was it about the material that best stopped the marble?

_____

_____

Name _____

Read the task. Then, follow the directions to complete the task.

Making It Rain

Make a rain stick. A rain stick is a tube filled with objects. When the tube is turned over, the objects move in the tube and make the sound of rain. Your tube's "rain" must last for at least 3 seconds.

Materials

cardboard tubes	beads
fabric scraps	pebbles
waxed paper	marbles
rubber bands	toothpicks
tape and glue	wooden skewers
dried beans	aluminum foil
rice	timer

Caution: Ensure that students are properly supervised while using sharp objects.

 Ask

What do you know? What do you need to know to get started?

_____

_____

 **Imagine**

What could you do?

_____

_____

 Plan

Choose an idea. Draw a model.

📝 Plan

What are your steps? Use your model to guide your plan.

_____

_____

_____

🛠️ Create

Follow your plan. What is working? Do you need to try something else?

_____

_____

🔄 Improve

How could you make it better?

_____

_____

💬 Communicate

How well did it work? Is the problem solved?

_____

_____

☀️ Reflect

How did the challenge of time affect the way you made your rain stick?

_____

_____

Name _____

Read the task. Then, follow the directions to complete the task.

A Wet "Suit"

Choose a card from the deck. Make a "wet suit" for your card so that it will stay dry after being dunked in water for 10 seconds.

Materials

deck of playing cards,
   construction paper
   squares, or thin
   cardboard
container of water
paper towels
copy paper
newspaper

waxed paper
aluminum foil
plastic wrap
felt
rubber bands
scissors
tape and glue
timer

 Ask

What do you know? What do you need to know to get started?

_____

_____

Imagine

What could you do?

_____

_____

Plan

Choose an idea. Draw a model.

Plan

What are your steps? Use your model to guide your plan.

_____

_____

_____

Create

Follow your plan. What is working? Do you need to try something else?

_____

_____

Improve

How could you make it better?

_____

_____

Communicate

How well did it work? Is the problem solved?

_____

_____

Reflect

How did what you know about the materials affect how you made your "wet suit"?

_____

_____

Name _____

Read the task. Then, follow the directions to complete the task.

A Sort of Color Sorting

Sort a set of colored objects while looking through colored lenses to find out what colors affect your ability to see color.

Materials

plastic toy eyeglasses, embroidery hoops, or cardboard frames

colored candies, cereal, or squares of paper

red, yellow, and blue cellophane

tape and glue

small containers for sorting (optional)

Caution: Before beginning any food activity, ask families' permission and inquire about students' food allergies and religious or other food restrictions.

Ask

What do you know? What do you need to know to get started?

_____

_____

Imagine

What could you do?

_____

_____

Plan

Choose an idea. Draw a model.

📓 Plan

What are your steps? Use your model to guide your plan.

_____

_____

🛠 Create

Follow your plan. What is working? Do you need to try something else?

_____

_____

🔄 Improve

How could you make it better?

_____

_____

💬 Communicate

How well did it work? Is the problem solved?

_____

_____

☀ Reflect

What color lenses were easiest to see the objects' colors through? What color lenses were most difficult? Why?

_____

_____

Name _____

Read the task. Then, follow the directions to complete the task.

Bag of Bananas

Use only 1 grocery bag to make a handled holder for 1 banana. You must be able to carry your banana 5 steps while only touching the handle of your holder.

Materials

paper grocery bag
    with bottom cut out
    or plastic grocery
    bag with bottom cut
    open

banana
scissors
tape

Caution: Before beginning any food activity, ask families' permission and inquire about students' food allergies and religious or other food restrictions.

Ask

What do you know? What do you need to know to get started?

_____

_____

Imagine

What could you do?

_____

_____

Plan

Choose an idea. Draw a model.

📝 Plan

What are your steps? Use your model to guide your plan.

_____

_____

_____

🛠 Create

Follow your plan. What is working? Do you need to try something else?

_____

_____

🔄 Improve

How could you make it better?

_____

_____

💬 Communicate

How well did it work? Is the problem solved?

_____

_____

☀ Reflect

How did the banana's shape affect how you made your holder?

_____

_____

Name _____

Read the task. Then, follow the directions to complete the task.

Tube Dude

Use only cardboard tubes to create a robot that is at least 6 tubes tall. The robot must have a body, arms, legs, and a head.

Materials

a variety of
   decorative objects,
   such as buttons,
   chenille stems,
   stickers, wiggly eyes

bathroom tissue tubes
scissors
glue

Ask

What do you know? What do you need to know to get started?

_____

_____

Imagine

What could you do?

_____

_____

Plan

Choose an idea. Draw a model.

📝 Plan

What are your steps? Use your model to guide your plan.

_____

_____

_____

🛠 Create

Follow your plan. What is working? Do you need to try something else?

_____

_____

🔄 Improve

How could you make it better?

_____

_____

💬 Communicate

How well did it work? Is the problem solved?

_____

_____

☀ Reflect

How did cutting into the tubes affect how the robot could be built?

_____

_____

Name _____

Read the task. Then, follow the directions to complete the task.

Command Me!

Write a list of commands for a friend to follow across a grid using a coin or other marker. Your list must have 15 commands. The coin or marker must stay within the borders of the grid. Example commands are *right 3 spaces*, *left 1 space*, *up 2 spaces*, *down 4 spaces*, etc.

Materials

pencil   paper
hundred board or   coin or other small
    other 10 x 10 grid   marker

🌓 Ask

What do you know? What do you need to know to get started?

_____

_____

💭 Imagine

What could you do?

_____

_____

📝 Plan

Choose an idea. Draw a model.

📝 Plan

What are your steps? Use your model to guide your plan.

_____

_____

_____

✂️ Create

Follow your plan. What is working? Do you need to try something else?

_____

_____

🔄 Improve

How could you make it better?

_____

_____

💬 Communicate

How well did it work? Is the problem solved?

_____

_____

☀️ Reflect

How did where the marker was placed at the beginning affect the set of commands?

_____

_____

Name _____

Read the task. Then, follow the directions to complete the task.

Loop-the-Loop

Using only paper loops, tape, and a drinking straw, make an airplane that can fly at least 3 feet (1 m).

Materials

paper strips in a   plastic drinking straw
   variety of widths   tape
   and lengths,   scissors
   pretaped into loops   yardstick (meterstick)

Ask

What do you know? What do you need to know to get started?

_____

_____

Imagine

What could you do?

_____

_____

Plan

Choose an idea. Draw a model.

📝 Plan

What are your steps? Use your model to guide your plan.

_____

_____

_____

⚒️ Create

Follow your plan. What is working? Do you need to try something else?

_____

_____

🔄 Improve

How could you make it better?

_____

_____

💬 Communicate

How well did it work? Is the problem solved?

_____

_____

☀️ Reflect

How did the length of the straw or the number of paper loops change how far your airplane flew?

_____

Name _____

Read the task. Then, follow the directions to complete the task.

Farmyard Fit

Build a model pen that will hold 8 farm animals. There must be a space the length of a paper clip between animals and between any animal and a side of the pen.

Materials

craft sticks
chenille stems
play dough
glue
tape

large paper clip
8 coins or linking
    cubes to represent
    farm animals

Ask

What do you know? What do you need to know to get started?

_____

_____

Imagine

What could you do?

_____

_____

Plan

Choose an idea. Draw a model.

Plan

What are your steps? Use your model to guide your plan.

_____

_____

_____

Create

Follow your plan. What is working? Do you need to try something else?

_____

_____

Improve

How could you make it better?

_____

_____

Communicate

How well did it work? Is the problem solved?

_____

_____

Reflect

What shape pen was easiest to make to complete the task?

_____

_____

Name _____

Read the task. Then, follow the directions to complete the task.

A Double Watering Can

Make a watering tool that can water at least 2 plants at one time.

Materials

water   tape
plastic water bottle   fake plants or empty
disposable foam cups   pots
toothpicks

Caution: Ensure that students are properly supervised while using sharp objects.

Ask

What do you know? What do you need to know to get started?

_____

_____

Imagine

What could you do?

_____

_____

Plan

Choose an idea. Draw a model.

📝 Plan

What are your steps? Use your model to guide your plan.

_____

_____

_____

🛠️ Create

Follow your plan. What is working? Do you need to try something else?

_____

_____

_____

🔄 Improve

How could you make it better?

_____

_____

_____

💬 Communicate

How well did it work? Is the problem solved?

_____

_____

_____

☀️ Reflect

To do this task, you had to think about the flow of water. How did this help you make a tool that could water 2 plants at one time?

_____

_____

_____

Name _____

Read the task. Then, follow the directions to complete the task.

Rough Road: Friction

Find the best material to stop a toy car traveling down a ramp.

Materials

wood board or heavy
   cardboard for the
   ramp
bubble packaging
   material
sandpaper
washcloth

aluminum foil
paper
waxed paper
one-sided corrugated
   cardboard
small toy car

 Ask

What do you know? What do you need to know to get started?

_____

_____

 Imagine

What could you do?

_____

_____

Plan

Choose an idea. Draw a model.

📝 Plan

What are your steps? Use your model to guide your plan.

_____

_____

_____

✖️ Create

Follow your plan. What is working? Do you need to try something else?

_____

_____

🔄 Improve

How could you make it better?

_____

_____

💬 Communicate

How well did it work? Is the problem solved?

_____

_____

☀️ Reflect

How did the material you chose affect the speed of the car? How might ramp height affect the speed of the car?

_____

_____

Name _____

Read the task. Then, follow the directions to complete the task.

Bright Baby Toy

Make two toys that a baby would enjoy looking at. Make one with water and the other with oil.

Materials

water
vegetable oil
clear plastic
   containers with lids
beads
colored rice

colored pasta
sequins and confetti
pom-poms
buttons
liquid food coloring
plastic spoons

Caution: Before beginning any food activity, ask families' permission and inquire about students' food allergies and religious or other food restrictions.

Caution: Before beginning this activity, ask families' permission and inquire about students' skin sensitivities and/or allergies.

Ask

What do you know? What do you need to know to get started?

_____

_____

Imagine

What could you do?

_____

_____

Plan

Choose an idea. Draw a model.

```



```

📓 Plan

What are your steps? Use your model to guide your plan.

_____

_____

_____

🛠️ Create

Follow your plan. What is working? Do you need to try something else?

_____

_____

🔄 Improve

How could you make it better?

_____

_____

💬 Communicate

How well did it work? Is the problem solved?

_____

_____

☀️ Reflect

Compare how materials move in oil to how they move in water.

_____

_____

Name _____

Read the task. Then, follow the directions to complete the task.

Marble Roll and Run

Make a maze in which a rolling marble changes direction at least 5 times.

Materials

marble   cardboard tubes
cardboard   paper
disposable foam   glue
  sheets   tape
wooden craft sticks

 Ask

What do you know? What do you need to know to get started?

_____

_____

Imagine

What could you do?

_____

_____

Plan

Choose an idea. Draw a model.

📝 Plan

What are your steps? Use your model to guide your plan.

_____

_____

_____

🔧 Create

Follow your plan. What is working? Do you need to try something else?

_____

_____

🔄 Improve

How could you make it better?

_____

_____

💬 Communicate

How well did it work? Is the problem solved?

_____

_____

☀️ Reflect

How did you position your materials so that the marble kept rolling as it changed directions?

_____

_____

Name _____

Read the task. Then, follow the directions to complete the task.

Coin Sifter

Create a sifter so that when coins are sifted, only quarters are left.

Materials

a variety of coins:
   pennies, nickels,
   dimes, quarters

cardboard box lid
scissors
pencil

❓ Ask

What do you know? What do you need to know to get started?

_____

_____

💭 Imagine

What could you do?

_____

_____

📝 Plan

Choose an idea. Draw a model.

Plan

What are your steps? Use your model to guide your plan.

_____

_____

_____

Create

Follow your plan. What is working? Do you need to try something else?

_____

_____

Improve

How could you make it better?

_____

_____

Communicate

How well did it work? Is the problem solved?

_____

_____

Reflect

How would your design change if you wanted to keep dimes? Would it work? Why or why not?

_____

_____

Name _____

Read the task. Then, follow the directions to complete the task.

Super Paper: Solid Shapes

Use only paper and tape to make a solid shape that will hold a board book at least 2 inches (5 cm) off the ground for at least 10 seconds.

Materials

board book	scissors
copy paper	timer
tape	ruler

🅱 Ask

What do you know? What do you need to know to get started?

_____

_____

💭 Imagine

What could you do?

_____

_____

📝 Plan

Choose an idea. Draw a model.

📝 Plan

What are your steps? Use your model to guide your plan.

_____

_____

_____

🛠 Create

Follow your plan. What is working? Do you need to try something else?

_____

_____

🔄 Improve

How could you make it better?

_____

_____

💬 Communicate

How well did it work? Is the problem solved?

_____

_____

☀ Reflect

What solid shape worked best? Name a solid shape that would not work. Why?

_____

_____

Name _____

Read the task. Then, follow the directions to complete the task.

"Ice" and Cozy: Insulation

Design a covering for an ice cube so that the ice stays frozen for 30 minutes at room temperature.

Materials

resealable bags	newspaper
ice cube	shortening
facial tissues	water
paper towels	tape and scissors
aluminum foil	plastic spoons
shredded paper	plastic plates or bowls

Caution: Before beginning any food activity, ask families' permission and inquire about students' food allergies and religious or other food restrictions.

Caution: Before beginning this activity, ask families' permission and inquire about students' skin sensitivities and/or allergies.

 Ask

What do you know? What do you need to know to get started?

_____

_____

 Imagine

What could you do?

_____

_____

Plan

Choose an idea. Draw a model.

Plan

What are your steps? Use your model to guide your plan.

_____

_____

_____

Create

Follow your plan. What is working? Do you need to try something else?

_____

_____

Improve

How could you make it better?

_____

_____

Communicate

How well did it work? Is the problem solved?

_____

_____

Reflect

How did this task help you understand how polar animals stay warm?

_____

_____

Name _____

Read the task. Then, follow the directions to complete the task.

Someone's Been Sitting in My Chair!

In the story "Goldilocks and the Three Bears," Baby Bear's chair breaks. Make a new chair for Baby Bear. Your chair must support the weight of a small teddy bear for 1 minute.

Materials

wooden craft sticks	cardboard tubes
plastic drinking straws	small teddy bear
a variety of empty	scissors
boxes including	tape
facial tissue boxes	glue
and cereal boxes	timer

Ask

What do you know? What do you need to know to get started?

_____

_____

Imagine

What could you do?

_____

_____

Plan

Choose an idea. Draw a model.

📓 Plan

What are your steps? Use your model to guide your plan.

_____

_____

🛠 Create

Follow your plan. What is working? Do you need to try something else?

_____

_____

🔄 Improve

How could you make it better?

_____

_____

💬 Communicate

How well did it work? Is the problem solved?

_____

_____

☀ Reflect

What parts of the chair were important to make strong? Why?

_____

_____

Name _____

Read the task. Then, follow the directions to complete the task.

Float-a-Boat: Buoyancy

Build a boat using only 1 square foot (929 sq cm) of aluminum foil. Your boat needs to stay afloat while holding the most pennies possible.

Materials

aluminum foil   pennies
large container of
   water

🌀 Ask

What do you know? What do you need to know to get started?

_____

_____

💭 Imagine

What could you do?

_____

_____

📝 Plan

Choose an idea. Draw a model.

📓 Plan

What are your steps? Use your model to guide your plan.

_____

_____

_____

🛠️ Create

Follow your plan. What is working? Do you need to try something else?

_____

_____

🔄 Improve

How could you make it better?

_____

_____

💬 Communicate

How well did it work? Is the problem solved?

_____

_____

🌟 Reflect

What if you used twice as much foil to make your boat? How many pennies do you think your boat would hold then?

_____

_____